How To Raise Strong & Healthy Pigs

I0489979

HTeBooks

Copyright © 2016

Disclaimer

This book is designed to provide condensed information. It is not intended to reprint all the information that is otherwise available, but instead to complement, amplify and supplement other texts. You are urged to read all the available material, learn as much as possible and tailor the information to your individual needs.

Every effort has been made to make this book as complete and as accurate as possible. However, there may be mistakes, both typographical and in content. Therefore, this text should be used only as a general guide and not as the ultimate source of information. The purpose of this book is to educate.

The author or the publisher shall have neither liability nor responsibility to any person or entity with respect to any loss or damage caused, or alleged to have been caused, directly or indirectly, by the information contained in this book.

Table of Contents

How Will This Book Help You?

For most people, the misconception that pigs are a dirty and difficult animal to rear has led them to settle for the cheap pork sold by meat processing companies. However, what most people are unaware of is that, healthy, pigs are one of the most social animals there is. If you are reading this book, then chances are you have already decided to rear pigs at your home, and are looking for important guidelines to carry you through. This book contains a simple step by step procedure, including the breeds of pigs available, food and water recommendations, as well as how to handle pests and diseases.

Once your pigs have matured, you will have several advantages, which include their meat either as pork, bacon, sausages or ham, all of which are a good source of good quality protein.

Starting Out

"I am fond of pigs. Dogs look up to us. Cats look down on us. Pigs treat us as equals."

- Winston S. Churchill

In this age of preservatives, overcrowded feedlots, hormones, antibiotics and quick cure methods, the only practical way of being assured of quality meat is producing your very own. Moreover, raising a pig is a project particularly suited for a beginning or small farmer for three main reasons. For starters, it has a low investment capital, it is a short-term project, and you can acquire a substantial amount of quality food from your family garden for the pigs at negligible cost.

How many pigs to raise

The truth is that a lone hog does not grow well at all, as he enjoys having company. The challenge is that the average family today does not actually eat more than one pig per year. There are two possible ways out of this dilemma. You could find a friend who is interested in raising a pig but does not have a place to do it or you could consider raising a second one for your comrade, who will then meet his share of the expenses and assist with the butchering.

It will be much less of a hustle to raise two pigs as compared to one; the hogs will have each other's company, and you will subsequently be doing someone a favor! The second option is to go ahead and raise both pigs, then butcher one and sell the other. Although there are laws in all states regarding the butchering and selling of meat, in general practice, you are less likely to be bothered as a small farmer who breeds for himself and sells to his friends. If you are in doubt about the feasibility of selling your home butchered meat, ask around for neighbors and friends who would be interested in purchasing fresh pork next fall. While fresh pork is not really that pricey, you should be able to get your returns, and a little extra to make up for the one you decide to keep.

Buying pigs

Young pigs in farm communities are usually offered for sale in the newspapers in the course of spring and the summer months. If none are listed in the local papers, check out the farm supply stores or visit a stock sale. However, avoid the temptation of buying your pig with the first sign of spring fever. For starters, you have probably not planted your garden yet, which means that it is no condition to feed a pig. Moreover, if you buy too early, you will end up with a costly, overgrown pet that is too expensive to maintain. You should know three things before you buy.

1: Pigs are nurtured and sold when they are 8 weeks old. Therefore, avoid buying pigs that are weaned at 6 weeks because they will not be as healthy.

2: It is generally accepted that a pig should be butchered when it is 6 months old so, if you continue feeding him after this age, you will only be investing money into the pig that no one will give you back.

3: Unless you have a walk in cooler at your disposal, you should not butcher the pig until frosty weather pulls in. The temperature range should be between 30 degrees and 40 degrees F in order for the meat to hang and cool after being butchered.

Good breeds to raise

The kunekune of New Zealand

Kunekune is a word used to mean fat and round. The kunekune pig is a small breed with a short snout and short legs. Some hang hanging tassels from the lower jaw. These pigs have varied colors and textures, and there is a clear distinction between their winter and summer coats. They feed on grass, and are not inclined to roam.

They also have an excellent ratio of fat to meat. This pig variety is sociable, and likes the company of humans, so you and your children can handle it safely, making it a very good pet.

*American mulefoot

This breed has a characteristic solid hoof, is hardy, fattens easily, and has a gentle disposition. At two years of age, this pig weighs 400 to 600 pounds.

*Berkshire or kurobuta

This is the oldest pig breed in Britain, and has its origins in the Berkshire County. It is a large breed, and is black in color, with white legs. This hog matures quickly.

*Red wattle

These rare hogs originated from New Caledonia, a French island, and were brought by the French to New Orleans in the late 1700s. You will find its excellent flavor resembling that of beef in taste and texture, and the pig has a lean and tender carcass. It is also a tasseled pig that adapts well to climates and is a brilliant forager. It weighs 600 to 800 pounds, but some can reach up to 1500 pounds.

*The Guinea Forest

These rare hogs are believed to have originated from the Guinea Coast in Africa and then traveled widely through the slave trade to England, Spain, France, and America. The breeds were a large and square breed with pointed ears and reddish bristly hair. These pigs are excellent foragers and hardy grazers that you can raise on pasture and still get pork and lard. Guinea hogs today are small, about 15 to 20 inches in height and 150 to 300 pounds in weight when fully grown, which makes them perfect for your small farm.

They have varied colors, but are often black and hairy. They are easy to care for and are very gentle, which makes them very popular with children.

*Tamworth

This is a small, red-coated, thrifty, rugged and active pig breed, whose origins are still unknown. It is however one of the oldest breeds, and is a descendant of native European pig stock and wild boars. Its ham is firm and muscular, but lacks the bulk and size found in several other breeds.

*Basque pigs

This is an endangered breed that once dominated the extreme southwestern Pyrenees. This French pork was ignored in favor of other breeds due to its high proportion of fat and low growth rate. After becoming almost extinct in the 70's, some breeders have since saved the breed in appreciation of raising their pigs in regards to ancestral traditions. The Basque pigs are a weather hardy breed that are alert, lively and great foragers in the wild for fruit, peas, acorns, chestnuts and grass. The breed is white with black spots, has extra tasty meat, but tends to grow slowly. They are good bacon and lard producers because of their ability to deposit fat easily.

*Cinta senese pigs

This is a very rare breed from the Tuscan native swine family, and is the only one that survived extinction. It is very resistant to bad weather, and is an almost savage breed. Therefore, it can be a secure food reserve for you and your children. However, this swine grows very slowly, which is why most farmers neglect them in favor of other breeds that grow faster. The cinta senese pigs are bred half-wild, eating roots, digging in the dirt and feeding on mushrooms. You can butcher the pig at two years of age. You may use its low fat, fragrant pork for cooking, but its main use is in the production of different types of tasty cold cuts.

Iberico pigs

The origin of this black breed of pigs can be traced back to ancient times, in Spain, Portugal and the southern and central territory of the Iberian Peninsula. Since these animals live freely, they are constantly on the move, which makes them burn more calories as compared to other species. The pig builds up fat between its muscular fibers and under its skin, which is the reason for the typical white streaks that are characteristic of the ham.

Meishan pig

This is a small to medium sized breed from the Taihu breed. It has black, wrinkled skin and large drooping ears. The breed is native to Southern china, and is most famous for its large litters of fifteen to sixteen piglets. These are perhaps one of the most prolific pig breeds in the world, and are known to produce two litters in a year. The USDA Agricultural Research Service to the United States imported this particular pig breed because of its fecundity. However, it lost flavor because of abundance of fat and its slow growth rate, although it matures early. While they are fat and slow growing, they do have a very good taste. They are also said to be resistant to certain diseases.

Large black

This breed originated from Chinese and found its way to England in the late 1800's. It is known for its taste, general hardiness, and pasture foraging skills. Large Blacks also feature short black hair, a long body and wide shoulders. When you butcher the pig, even at two hundred pounds, the excellent bellies, short muscle fibers, and micro marbling all create moist meat and exceptional bacon with a tasty flavor. The color of its coat makes it resistant to several sun borne conditions and its grazing ability and hardiness make it a sufficient meat producer. These pigs are very docile in nature and tend to move more deliberately and slowly than other breeds. The Large Blacks are also slow maturing and are known for their prolificacy, milk capacity and mothering ability. The sows bear 8 to 10 piglets.

Ossabaw Island, Georgia pig

This endangered breed originated from Spain and lives off the coast of Georgia. These pigs have a long snout and heavy coat. Their popularity within the chef community has been boosted by their fat quality and marbling. They are small and weigh less than 200 pounds and are less than two hundred inches tall at maturity, although they grow much larger in captivity. Its meat is considered a heritage product that is specially suited for use in whole pig roasts, cured meats, and in pork.

*Key point/action step

When looking for your own pigs, ask for either barrows (the castrated males), or sows (the females), because meat from an uncastrated male (boar), has a very unpleasant taste and odor. If possible, go for the huskiest looking hogs of the litter; the plump-looking hams that have short legs. You will need a box or wooden crate to transport them in since a pig does not handle like other animals. Ensure that you choose a suitable breed that you will enjoy having rather than one that will disappoint you.

Housing Your Pigs

"Quality, affordable housing is a key element of a strong and secure Iowa."

- Thomas Vilsack

Once you have determined the kinds of pigs you want to keep, the next step is to design a house to shelter your animals. The best way to house your pigs is to provide a shelter that can keep them comfortable in different seasons. The ideal ones are built from either bricks or concrete to make it easier to clean. However, you can still make your shelter using wood, and then surround it with a wire mesh. The work of the mesh is to prevent the pigs from nibbling the wood.

In addition to the structure, it is also important for the pigs to have access to shallow ponds or mud. This will come in handy during hot days, when your pigs need to cool off. You must also ensure that food and water are readily available when raising swines. However, since pigs are regarded as strong creatures, you need to secure them in strong enclosures at all times. The optimum length of a typical pig house is roughly 4 feet tall. The average weight of a typical swine is roughly 150 pounds, but some can weigh up to 800 pounds. Therefore, 8x6 ft area for each pig should be sufficient enough to allow them to move around the shelter with ease.

When constructing a pig house, it is crucial that you get a good guide and execute your ideal shelter very well in order to avoid extra expenses caused by repairs. The materials needed are usually concrete or bricks, with wire mesh, six metal posts, straw, plywood, and hog panels. The roof is normally made from wood. To achieve good ventilation, raise the roof to 12 inches above the wall. For adequate bedding, you may want to place straws all over the floor. Moreover, it is also advisable to plan the location wisely. It should be within an area that has enough water supplies such as an outdoor well or a pond. Some farmers find it useful to dig about a foot or so into the ground before building the house. This is especially beneficial in cooling and insulating the inside of the pig house in order to keep them comfortable all year round. Generally, it is not that difficult to obtain a pig shelter, as long as your objective is to provide a comfortable dwelling for your pigs.

Shelter requirements

The temperature range needed to achieve maximum pig productivity is known as the thermoneutral zone. The heat production of pigs within this zone does not depend on air temperature, and is therefore determined by its food intake and live weight. Critical temperatures will vary depending on the specific conditions in your piggery and your pigs' total weight. However, if they spend more time shivering or huddling, and eat more than normal, they are usually cold. On the other hand, if they avoid body contact with their counterpart mates, foul clean areas of their pens, eat less and pant at more than fifty breaths per minute, then they will become warm. The highest tolerable thermoneutral zone temperature is between 6 to 8 degrees Celsius, beyond which serious problems are likely to set in.

If the temperature in the immediate surroundings of your pigs falls below the lower critical temperature, which is the lowest tolerable temperature, your pig will be inclined to maintain its body heat using some of its energy. Older pigs can withstand lower temperatures for short periods without signs of obvious ill health, at the expense of the efficiency of their food conversion. It is also important to note that the LCT decreases with the age of the pig. The ideal temperature for newborn piglets is usually between 27 and 35 degrees Celsius. However, their ability to withstand cold temperature becomes limited, as the piglets grow older. You are more likely to lose some piglets very quickly if the microclimate stays below 16 degrees Celsius. Fatal chilling will take place within minutes at temperatures below 2 degrees Celsius, unless you provide them with warmth.

However, if the area is not draughty, your pigs can withstand low temperature. Some areas to keep an eye on are open-ended trenches that allow draught through the slats, wall cracks or near floor level, as well as uncovered heat lamps in cold buildings that can result in a draught when hot air is displaced by cold air at the floor level. In addition, be sure to use creep boxes or covers to reduce draughts and retain warmth.

It is usually quite easy to warm a dry concrete floor for your pigs. However, while concrete tends to retain heat considerably well, it also increases the harmful low temperature effects when damp. Your pig will pass considerable heat into the damp concrete floors, despite the air temperature being reasonable.

Untreated wood shavings or dry straws are excellent insulations for very young pigs against cold conditions. If the temperature in the immediate surroundings of your pigs' shelter goes beyond the UCT, it will make the pig severely distressed. As the pig ages, the UCT reduces. Young pigs are the most susceptible to cold, while larger and older animals are more vulnerable to rising temperatures. In fact, temperatures above 27 degrees C are widely considered undesirable for breeders, finishers, and growers. However, if your pigs have sufficient air movement in the shelter, you can reduce heat stress in dry climates through spray or dip cooling. The resulting water evaporation from the skin of the pigs can significantly reduce excessive body heat.

Regardless of environmental conditions, you need to provide a minimum amount of fresh air to your pig house in order to eliminate odors, bacteria, airborne dust, ammonia, carbon dioxide, and water vapor. However, ventilation tends to reduce the temperature in the shelter, which is why it is important to insulate the walls and roof to minimize heat gain or loss through conduction, and keep it draught proof to limit uncontrolled air change. If your insulation does not create its own vapor barrier, you can protect it with a vapor barrier to reduce condensation within the shed. This serves to protect the interior linings, as well as reducing the amount of ventilation you will need to prevent condensation. Be sure to direct cold ventilating air such that it creates air circulation without flowing directly onto the animals within the pig house. A conventionally and naturally ventilated pig house normally involves using side wall vents together with a ridge vent.

Building orientation

If the long axis spreads from east to west, long and narrow buildings tend to be warmer in winter and cooler in summers. You should situate your pig shed to make the most of prevailing winds for coolness in the summer months. Conversely, you should protect ventilation openings in winter against prevailing winds. You can achieve this by planting trees in a shelterbelt that will not impede with the airflow needed for cooling during summer. These will soften the visual impact and enhance the appearance of your piggery.

*Key point/action step

Maintaining adequate temperature in the pig's shed is very important for the survival of your pigs. Your goal is to ensure that the shed is not too hot by providing adequate ventilation while also making sure that it is not too cold by providing the necessary insulation. Also, have in mind that the age of the pig determines the kind of house they need. While a large pig is susceptible to high temperatures, a young one is more susceptible to cold temperatures so have this in mind when building a pig shed.

Feeding Your Pigs

"Animals feed; man eats; only a man of wit knows how to eat."

- Jean-Anthelme Brillat-Savarin

Pigs are omnivorous, just like us, and can eat a wide range of foods. They also need a balanced diet like us including fiber, proteins, energy, vitamins, and minerals to grow strong and healthy. In keeping pigs, feed will be your biggest expenditure, so it is useful to get it right. Pigs require a constant supply of fresh drinking water, like all the other animals. Pigs tend to tip the trough to make a wallow, and are known to stand in the water trough and wash themselves in it, so you have to check, clean and refill the trough on a regular basis.

Galvanized troughs are strong and easy to clean, as well as hard for the swine to tip over. You can also get automatic drinkers, but these are not as much enjoyable for the pigs. If you have the right equipment and knowledge, you can mix your own food, but most people prefer using commercial pig food. There are several different feed producers, including both GM feeds and organic feeds. Be sure to find a dry and rodent free place to store the bags. Store the food in a rodent proof container once you open the bags. In addition, make sure you sweep out and get rid of any spillages. Check regularly to ensure that the bags have not been damaged by rodents, and if they have, deal with the damage right away. Always check the expiry date before purchasing your feed. It may retain the oil, fiber and protein value, but its mineral and vitamin level will reduce after it.

Pigs are usually fed twice per day. However, the amount of feed will depend on the reproductive state and age of your pigs. A foraging pig will get some of its food from natural sources, if the foraging area can be able to provide it. This would include apples, acorns, brambles, grass and, yes, earthworms. It is crucial that you supplement these with an all-rounded compound feed so that your pigs receive all the essential nutrients they need. However, foods such as carrots and potatoes do not cater for the nutritious value of compound feeds, and should therefore not be fed to the pigs. In fact, it is illegal to feed pigs with any household waste. You can, however, feed them with vegetables and fruits from non catering premises.

Pigs like to eat their feed when wet, so it is advisable to add surplus goat's milk or water to their feed, so long as the milk has not passed through the kitchen, after which it would be considered catering waste. In addition, unless you are registered with Animal Health, ensure that your pigs' diet does not consist of more than 80 percent of waste milk. Pig troughs are especially useful for sheds with more than one pig, as this will ensure that even the more timid members get enough to eat, or alternatively spread the feed around the ground widely. The latter should be done only on clean areas, but will inevitably lead to more feed waste.

Gilts

The gilts (young sows), will require 5 ½ lb (2.5 kg) of meal, cakes or sow breeders pencil per day. You should keep these until right before farrowing. However, be careful, keeping in mind that a maiden gilt is still in its growing stages, and needs to feed her unborn child. This means that she needs sufficient feed to be able to perform both functions. You should increase the rations of the gilt gradually to 4 kg after three weeks. This will help maintain her best. During gestation and after service, you may reduce the feed to 5 lb (2.4 kg) per day.

After farrowing

As soon as the pig has given birth to her litter, you must get her enough food for her to stay healthy and to provide sufficient quality milk for her piglets. Be sure to feed the sow roughly 6 lb or 3kg of meal, cakes or sow breeders pencils per day. If the sow has given birth to more than 6 piglets, you should feed her with an additional 0.5kg per each extra piglet. You can reduce this to 1.5 to 2 kg after weaning.

*Key point/action step

Proper feeding of the pig will determine if you will have healthy pigs or not. Remember that pigs also need clean water too so keep checking on the water in the trough as pigs are known to tip the trough and even step on the trough. If you have a pig with piglets, ensure that you provide more food so that the pig can provide adequate and quality milk for the piglets.

Hygiene

"Civilization is the distance that man has placed between himself and his own excreta."

- Brian W.Aldiss

As a pig farmer, you need to take precautionary measures to protect the health of your animals. Maintaining good health is not only vital to ensure acceptable standards of animal welfare, but also to maximize on their productivity.

External measures to prevent pathogens

*Your pig farm should be located far enough from any other farms, especially in the direction of the dominant wind. Flies can spread some infectious particles over short distances. Rodents too tend to spread diseases from one farm to the next.

*Purchasing policy: Purchasing should be limited and selected carefully on vaccination and sanitary status, when it comes to a breeding farm. Minimize purchasing sources in fattening farms.

*Quarantine: Once you bring in new animals into the farm, you should keep them in quarantine for veterinary observation. In fact, this is a legal obligation in several European countries. The practical period is usually 4 to 6 weeks.

*Minimize visitors. Make sure that anyone entering the farm is wearing boots and an overall. They should also wash and disinfect their hands before and after leaving the shed.

*Vehicles: In order to prevent cross-contamination, vehicles transporting animals are required to be cleaned thoroughly and disinfected after every trip. The golden rule is to clean first and then disinfect. A slightly alkaline detergent should be used for cleaning, as this will remove organic dirt such as fat and various proteins. Acidic products remove inorganic dirt such as lime scale. However, an excessively alkaline solution or one that contains chlorine or sodium hydroxide will corrode the truck, especially the parts made of aluminum. It is not possible to remove these specific kinds of dirt by using high pressure water alone. If possible, ensure that tracks that are not disinfected do not enter your farm for pick up, if you are raising several pigs for sale.

*Footbaths: Install a footbath at the entrance of every house. Make sure that the boots are properly cleaned before dipping them into the foot dip. In addition, the disinfectant should be powerful enough to kill all microorganisms. Ideally, you should renew it daily.

Internal measures to deal with pathogens within the farm

*Cleaning and disinfection: It is crucial to employ cleaning, disinfection and sanitary measures after every cycle to prevent development and progression of pathogens within your farm.

*Key point/action step

You want healthy pigs not only for the good of the pigs but also for your good and for that of your family. You would not want to have sickly pigs that can easily transfer pathogens to the home when you are in contact with them. Therefore, ensure that you emphasize on cleanliness.

How To Recognize Disease In Your Pig Farm

"He who cures a disease may be the skillfullest, but he that prevents it is the safest physician."

- Thomas Fuller

The first priority when it comes to managing disease in your pig farm is early recognition through using the senses of smell, touch, sound and sight to detect the sick animal and to distinguish it from the healthy animals. If possible, it is advisable to carry out a clinical examination of all your pigs every day.

The use of sight

*Lack of appetite is usually normal when an animal is bred in isolation, such as a confined sow. However, this is relatively hard to detect in group housed animals. If you notice a drop in feed intake or failure to eat by otherwise normal pigs, be alert at once and check for availability of water. Loss of appetite in all pigs in a group could highly be due to lack of water. Look for signs of disease if the water supply is okay.

*A dull appearance of listlessness is also an early sign of illness.

*If you notice rising of hair over one of your pig's body or shivering, take action at once as this is a common symptom of disease, and is an early sign of joint infections or streptococcal meningitis. If a pig is shivering with its hair on end as it lies on its belly, it could mean that the animal is either lame or scoured from a generalized septicemia, which is a bacteria in the bloodstream.

*Loss of weight in one of your pigs is a first sign of dehydration or loss of appetite due to pneumonia or diarrhea.

*Discharges from the eyes or nose mean that your pig has an upper respiratory infection. If it has excess salivation from its mouth, it could indicate an exotic disease like vesicular disease. A discharge from the vulva of sows could indicate endometritis, pyelonephritis, cystitis, or vaginitis.

*There is a wide range of diseases that could cause fecal changes, but sloppy feces could actually be normal. Watch out for signs of blood or mucus indicative of salmonella infections, swine dysentery, proliferative haemorrhagic enteropathy or gastric ulceration.

*If you notice vomiting from one of your pigs, it could be a sign of such diseases as transmissible gastro enteritis, or gastric ulceration in individual pigs.

*Skin changes can help you identify diseases, typified by chronic or acute lesions of lice and mange, although the former are now quite uncommon. Extreme bluing could indicate acute bacterial septicemia, acute viral infections, or a toxic condition, such as seen in mastitis, PRRS infections and flu.

*Respiration rates: If you have identified any of the mentioned changes, observe the rest of the pigs and compare the rate of respiration of both the suspect pigs and the normal ones. Determine if the breathing is deep with considerable chest movement because of consolidation of the lungs, and scarcity of oxygen, or if it is a very shallow breathing indicating pain and pleurisy.

Observing the group

You should set daily and regular time to examine all your pigs. You should also assess the environment of the shed by noting the following:

*Humidity

*Smell

*Ammonia levels as seen through breathing and the effect on eyes

*Temperature

*Ventilation

*Pig behavior

*Appetite

*Human reaction

*Abnormal changes in slurry and bedding

Changes in behavior

Pigs are known to be social animals that prefer being part of a group in a healthy condition. However, when ill, your pig will tend to rest on its own, and can even be isolated by the other members to the extent of being attacked. You should regard abnormal lying patterns with suspicion. Huddling, on the other hand, is common where several pigs are sick or if the environment is inadequate. Be alert if you notice reluctance in your pigs to show interest or rise in the presence of an observer.

*Key point/action step

No one knows your pigs better than you do. Therefore, ensure that you note any change of behavior as this could mean that they are unwell or are uncomfortable.

How to Apply What You've Learned?

Consumers have for long been happy to purchase low quality pork without giving much thought as to how it was produced. In truth, most of us have been unaware of the changes in the housing done at the pig industry. However, you may have noticed that the quality of the pork had changed and that it had become bland and dry. The use of words such as "quality assured", "all natural", and "grain fed" have all been served to keep you in the dark just that bit longer. When all is said and done, in order to raise strong and healthy pigs, you will need good management skills to produce a desirable product, and follow at least these five principles:

*Space: Sufficient room for the pig to escape any confrontation and behave naturally

*Diet: A well balanced diet with all the nutrients essential for your class of pig

*Water: A ready supply of fresh and clean water, or else your pigs will refuse to eat

*Shelter: Appropriate space with enough protection from environmental elements

*Stockmanship: Pigs that are handled well are happy pigs!